双语版·全4册

GARDEN BIRD

亲亲动物

如果麻雀来我家

[英]弗朗西斯·罗杰斯 [英]本·克里斯戴尔 著绘 范晓星 译 朱朝东 丁亮 审校

中信出版集团 | 北京

图书在版编目（CIP）数据

如果麻雀来我家：汉文、英文 / (英) 弗朗西斯·罗杰斯, (英) 本克里斯戴尔著绘；范晓星译. -- 北京：中信出版社, 2023.3
（DK亲亲动物：双语版：全4册）
ISBN 978-7-5217-5239-7

Ⅰ. ①如… Ⅱ. ①弗… ②本… ③范… Ⅲ. ①麻雀－少儿读物－汉、英 Ⅳ. ①Q959.7-49

中国国家版本馆CIP数据核字（2023）第021871号

Original Title: How can I help Rory the garden bird?

如果麻雀来我家
（DK 亲亲动物双语版　全 4 册）

著　　绘：［英］弗朗西斯・罗杰斯　［英］本・克里斯戴尔
译　　者：范晓星
出版发行：中信出版集团股份有限公司
（北京市朝阳区东三环北路 27 号嘉铭中心　邮编　100020）
承 印 者：北京顶佳世纪印刷有限公司

开　　本：889mm × 1194mm　1/20　　印　　张：2　　字　　数：115 千字
版　　次：2023 年 3 月第 1 版　　印　　次：2023 年 3 月第 1 次印刷
京权图字：01-2022-4478　　审 图 号：GS 京（2022）1525号（本书插图系原文插图）
书　　号：ISBN 978-7-5217-5239-7
定　　价：156.00 元（全 4 册）

出　　品　中信儿童书店
图书策划　红披风
策划编辑　陈瑜
责任编辑　袁慧
营销编辑　易晓倩　李鑫橦　高铭霞
装帧设计　哈_哈

服务热线：400-600-8099
投稿邮箱：author@citicpub.com

致所有好奇的孩子！

www.dk.com

混合产品
纸张 | 支持负责任林业
FSC® C018179

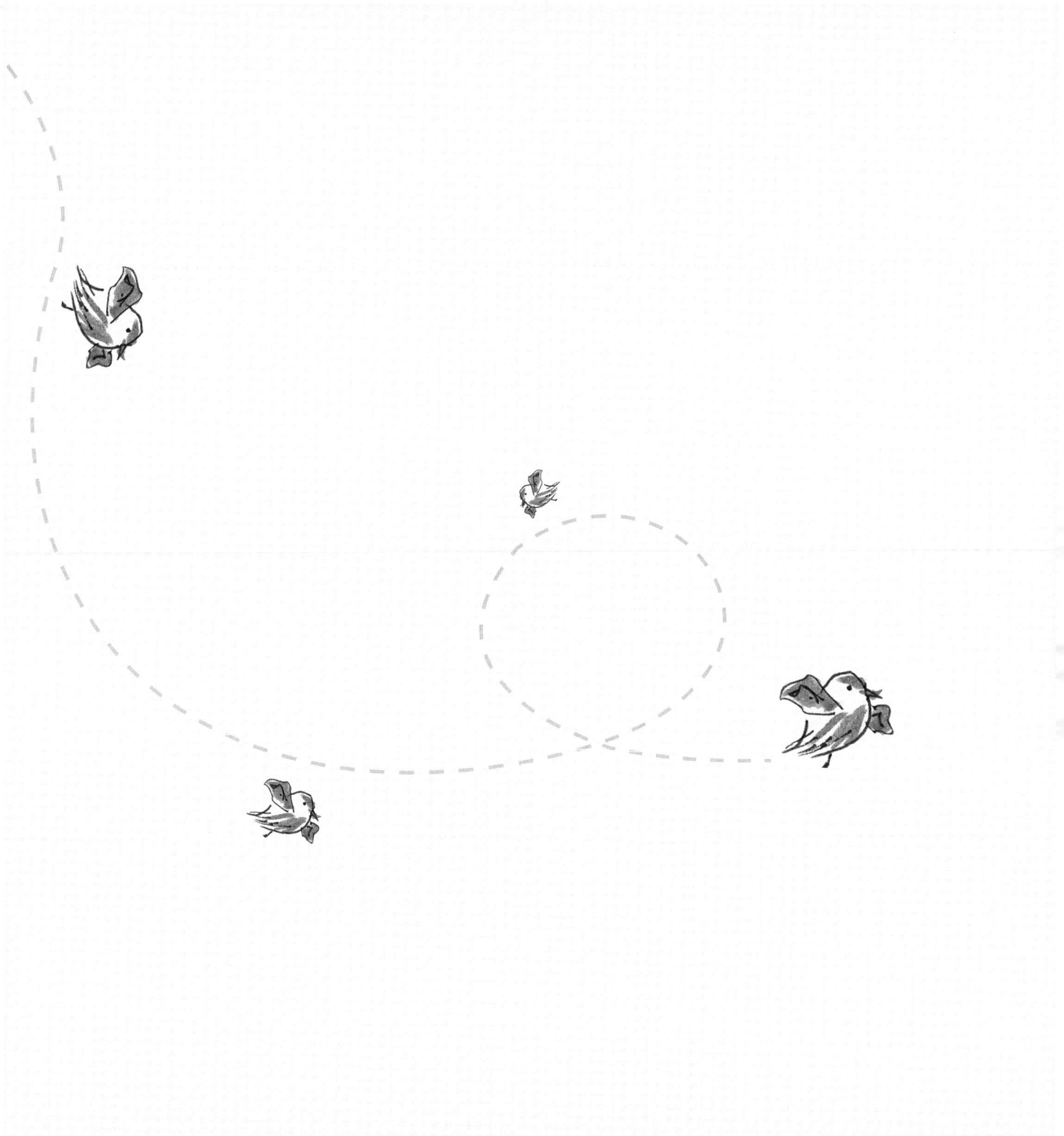

你好，我叫罗芮。

我是一只花园鸟[①]。

我想请你帮帮忙，可以吗？

Hello, my name is Rory.

I am a garden bird and I need your help.

---

① 花园鸟是经常在人类建造的花园区域活动的鸟类，麻雀是其中的一种。

我能到你家花园做客吗？
只要一点点食物和水就好。

Let me visit your garden
for food and water.

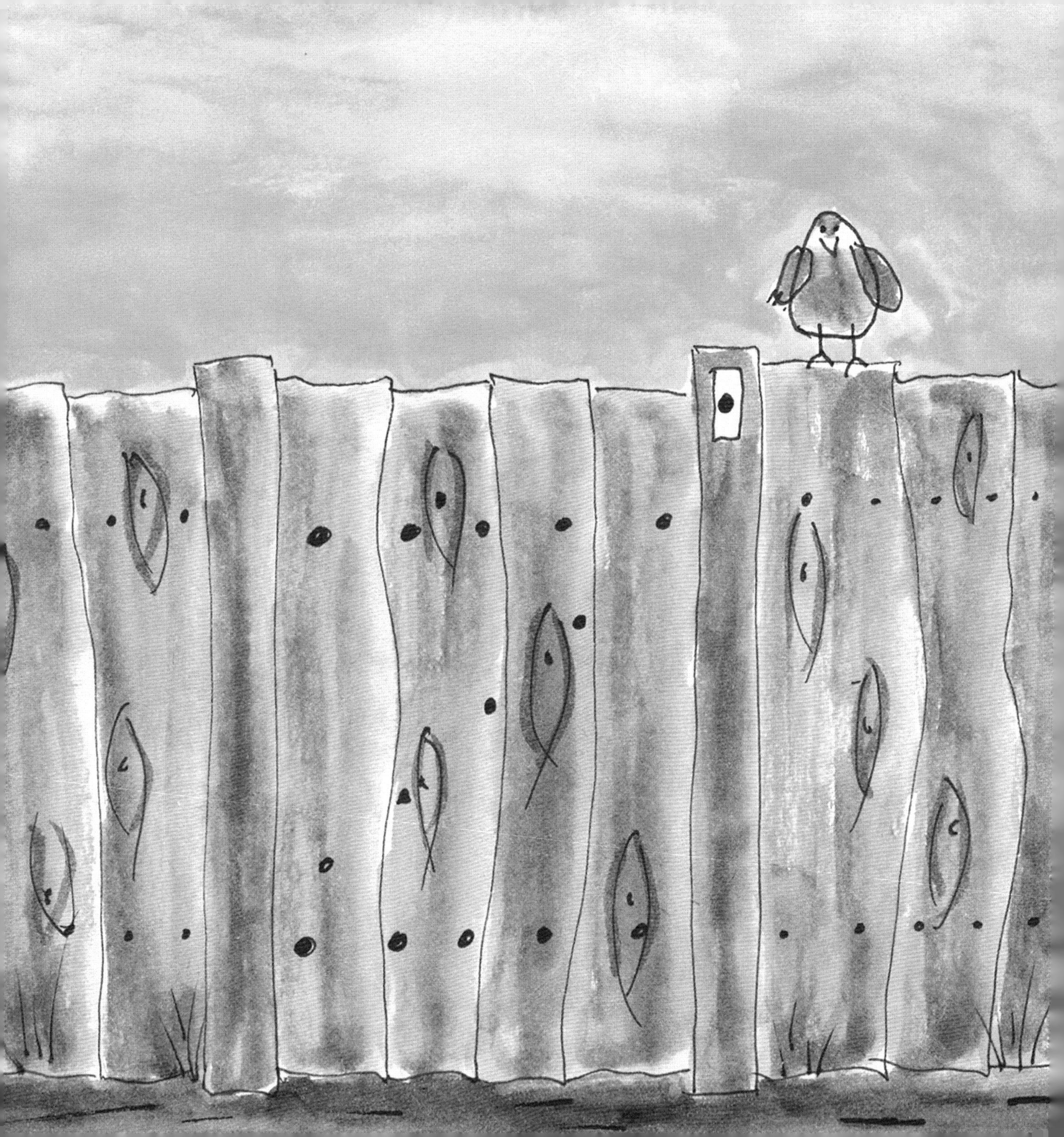

我想喝点儿水，
还想洗个澡。

I like to drink and
wash in water.

给我准备一个鸟儿的洗澡盆吧。

Please give me a bird bath.

别忘了在里面加满
干净的水哟。
Keep water full and fresh.

夏天，我喜欢吃种子一类的鸟食。

I like to eat bird seed
in the summer.

冬天，可以把食物团成团儿喂给我。

I like to eat balls of
food in the winter.

如果洗澡盆里的水冻住了，
别忘了把冰化开哟。
Melt water if frozen.

我也喜欢吃甲虫和其他飞虫，
所以请你种些开花的植物，
吸引这些虫虫吧。

I also like to eat bugs and flies,
so please plant flowers to attract them.

树上是挂喂鸟器的好地方，
我在树上也会感觉很安全。

Trees are good for food
and I feel safe in them.

请在你家的树上，
为我安一个家。

Please give me a home
in your trees.

花园里如果有垃圾的话，
可能会给我带来危险。

Rubbish in your garden
can be a danger to me.

请保持花园干净整洁吧。

Please keep your garden tidy.

请不要在花园里
放这样的网。

Please do not use nets
in your garden.

我可能会被困在里面。

I can get stuck in them.

请经常清洗喂鸟器和洗澡盆，
不然我会生病的。

Please keep your bird feeders and bath clean
to stop me getting poorly.

热 水
Hot water

谢谢你为了保护我们做的一切。

Thank you for all your help.

我们为什么要保护家麻雀？

Why do we need to protect house sparrows?

罗芮是一种叫家麻雀的鸟。家麻雀生活在喧闹的城镇，它们会自己筑巢，也会在树丛、洞，甚至建筑物的缝隙里安家。

Rory is a type of bird called a house sparrow. House sparrows live in busy places like towns and cities, in nests, bushes, holes, and even in gaps in buildings.

遗憾的是，这些家麻雀面临着生存危机。也就是说，可能有一天，我们再也见不到它们了。

Unfortunately, these sparrows are endangered, which means that, one day, we may not see them any more.

不光是家麻雀有危险，其他种类的麻雀也都遇到了麻烦，所以我们要尽力去帮助这些小小的鸟！

It is not just house sparrows that are at risk, but other types of sparrows are also in trouble. That is why we all need to do what we can to help these little animals!

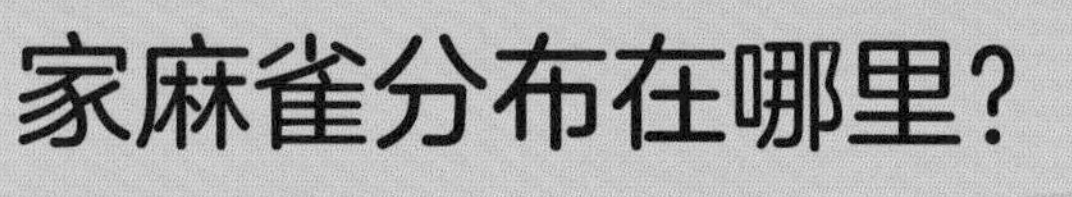

家麻雀分布在哪里?

Where in the world?

人类最早在欧洲、亚洲部分地区和北非发现家麻雀，后来，它们被带到了世界各地。这些长羽毛的小动物现在遍布全球大陆，除了南极洲!

House sparrows were first found in Europe and parts of Asia and northern Africa, but they have since been introduced to other parts of the world. These feathered friends can now be found on every continent except Antarctica!

家麻雀从喙到尾巴能有 15 厘米长，大概是一张钞票那么长。

A house sparrow can be up to 15 cm (6 in) from beak to tail-that is roughly the same length as a bank note!

你住的地方能看到家麻雀吗？
Can you spot house sparrows where you live?
它们展开翅膀后，翅膀尖之间的距离有 21~25 厘米，大概是一个足球的直径那么长！
When open, its wings can be 21-25 cm (8-10 in) long-that is around the size of a football!
雌性家麻雀
Female sparrow
有黑色或棕色的、短短的、粗粗的喙。
A black or brown and short and chunky beak.
全身有棕色的羽毛。
Brown feathers all over.
肚子上的羽毛是浅棕色的。
Pale brown colouring on its tummy.

全世界的花园鸟

Garden birds around the world

世界上有许多各种各样的花园鸟，也包括家麻雀。这里有……

There are lots of different garden birds in the world, as well as the house sparrow. Here are just a few that can be found…

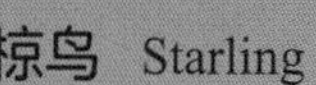

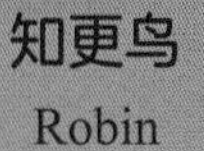

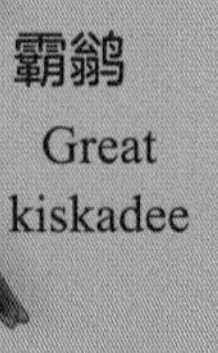

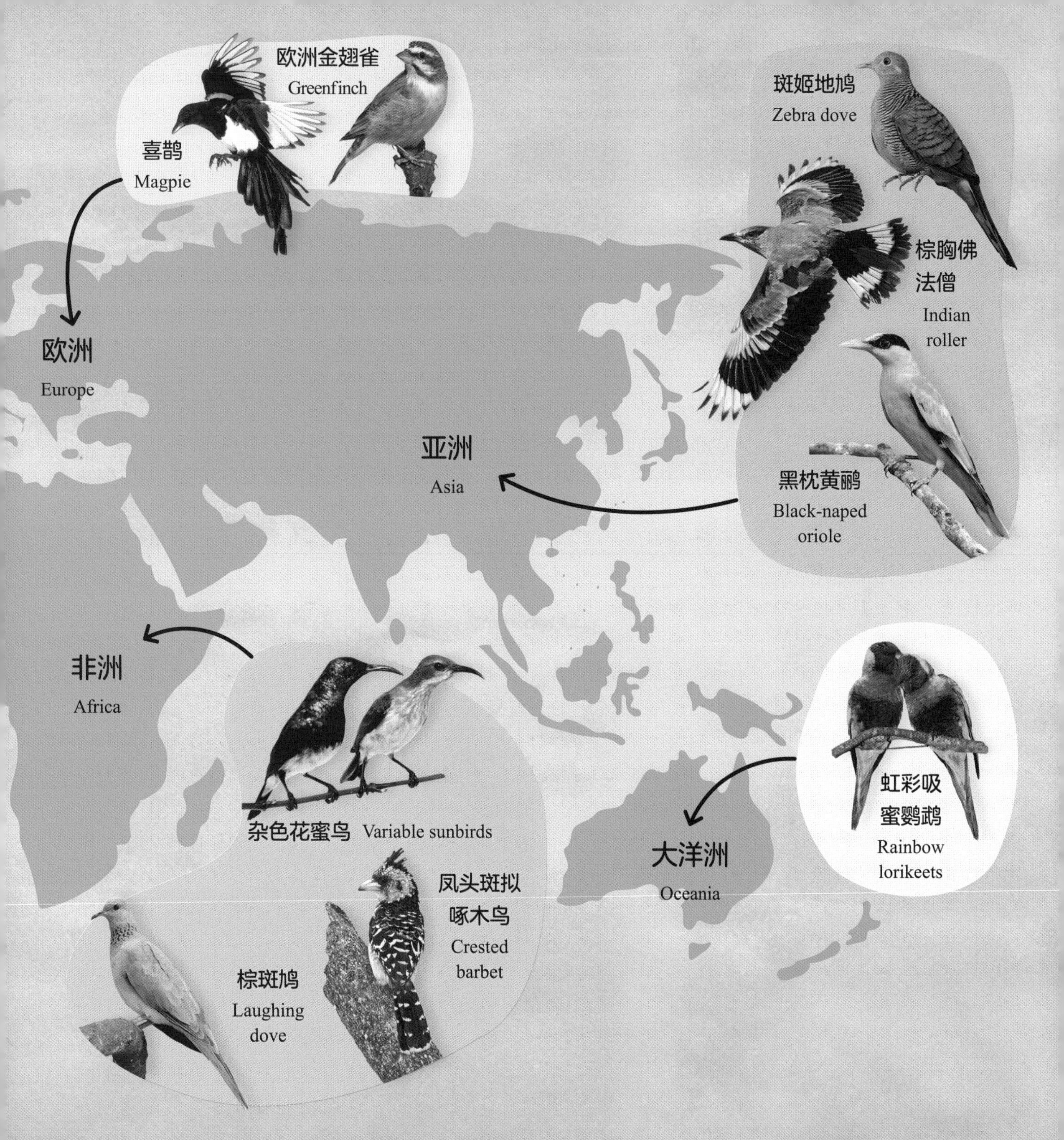
欧洲金翅雀
Greenfinch
喜鹊
Magpie
欧洲
Europe
斑姬地鸠
Zebra dove
棕胸佛
法僧
Indian
roller
亚洲
Asia
黑枕黄鹂
Black-naped
oriole
非洲
Africa
杂色花蜜鸟 Variable sunbirds
凤头斑拟
啄木鸟
Crested
barbet
棕斑鸠
Laughing
dove
大洋洲
Oceania
虹彩吸
蜜鹦鹉
Rainbow
lorikeets

鸣 谢

本出版社感谢以下机构提供照片使用权：

(a= 上方；b= 下方；c= 中间；f= 底图；l= 左侧；r= 右侧；t= 顶端)

34 Getty Images: James Warwick (tr, cl/x2, cr). **35 Chris Gomersall Photography: (cra). Dreamstime.com:** Victortyakht (tl); Wildlife World (clb). **36 Getty Images / iStock:** Dgwildlife (c). **37 Dreamstime.com:** Isselee (c). **38 Dorling Kindersley:** Alan Murphy (bc). **Dreamstime.com:** Steve Byland (cl); Svetlana Foote (ca); Tony Northrup / Acanonguy (cb); Vasyl Helevachuk (cr). **Getty Images:** Stockbyte / John Foxx (cra). **39 Dreamstime.com:** Peter Betts (bc); PeterWaters (crb); Davemontreuil (bl). **Getty Images / iStock:** PrinPrince (cr).

关于作者和绘者

本和弗朗西斯是一对夫妻，住在英国泰恩河畔纽卡斯尔。他们倾心于帮助家中花园里的野生动物。仲夏的夜晚，弗朗西斯醒来，有了一个灵感——创作鼓励小朋友加入他们的行动的系列童书。

弗朗西斯创作故事，本画插图，他们引领小读者进入一个栩栩如生的野生动物世界，快来欢迎花园里的小客人：小刺猬罗里、小麻雀罗芮、小蜜蜂罗丝、小蝴蝶罗克西。